Kay

DISCARD

Kw)

JUNGLES AND FORESTS

Design	David West Children's Book Design
Designer	Keith Newell
Editorial Planning	Clark Robinson Limited
Editor	Yvonne Ibazebo
Picture Researcher	Emma Krikler
Illustrators	Mike Saunders and Simon Tegg
Consultant	David Flint University lecturer

© Aladdin Books Ltd 1992

First published in
the United States in 1993 by
Gloucester Press
95 Madison Avenue
New York, NY 10016

Library of Congress Cataloging-in-Publication Data

Twist, Clint.
　　Jungles and forests / Clint Twist.
　　　　p.　　cm. — (Hands on science)
　　Includes index.
　　Summary: Examines the climatic conditions in a rain forest and their effects on the variety of plant and animal life found there. Includes related projects.
　　ISBN 0-531-17397-6
　　1. Rain forests—Juvenile literature. 2. Rain forest fauna—Juvenile literature. 3. Rain forest plants—Juvenile literature. 4. Rain forest ecology—Juvenile literature. [1. Rain forests. 2. Rain forest ecology—Experiments. 3. Experiments.] I. Title. II. Series.
QH86.T85　　1993
574.5'2642—dc20　　　　92-33916　　　CIP　　AC

All rights reserved

Printed in Belgium

HANDS·ON·SCIENCE

JUNGLES AND FORESTS

CLINT TWIST

GLOUCESTER PRESS
New York · London · Toronto · Sydney

CONTENTS

FOREST ZONES	6
NATURAL COVER	8
TROPICAL RAIN FORESTS	10
LAYERED FORESTS	12
DIVERSITY	14
HOW IT WORKS	16
TEMPERATE FORESTS	18
SPECIES AND SOIL	20
BOREAL FORESTS	22
WINTER SURVIVAL	24
OTHER FORESTS	26
FOREST MANAGEMENT	28
MYTHS OF THE FOREST	30
GLOSSARY	31
INDEX	32

This book is all about jungles and forests. The topics covered range from the different forest zones that occur around the world, such as rain forests and boreal forests, to the type of plants and animals that live in the forests. The book talks about the conditions that particular forests need in order to survive, and how human activity is threatening areas that were once forested. There are "hands on" projects for you to do that use everyday items as equipment. There are also "did you know" panels of information for added interest.

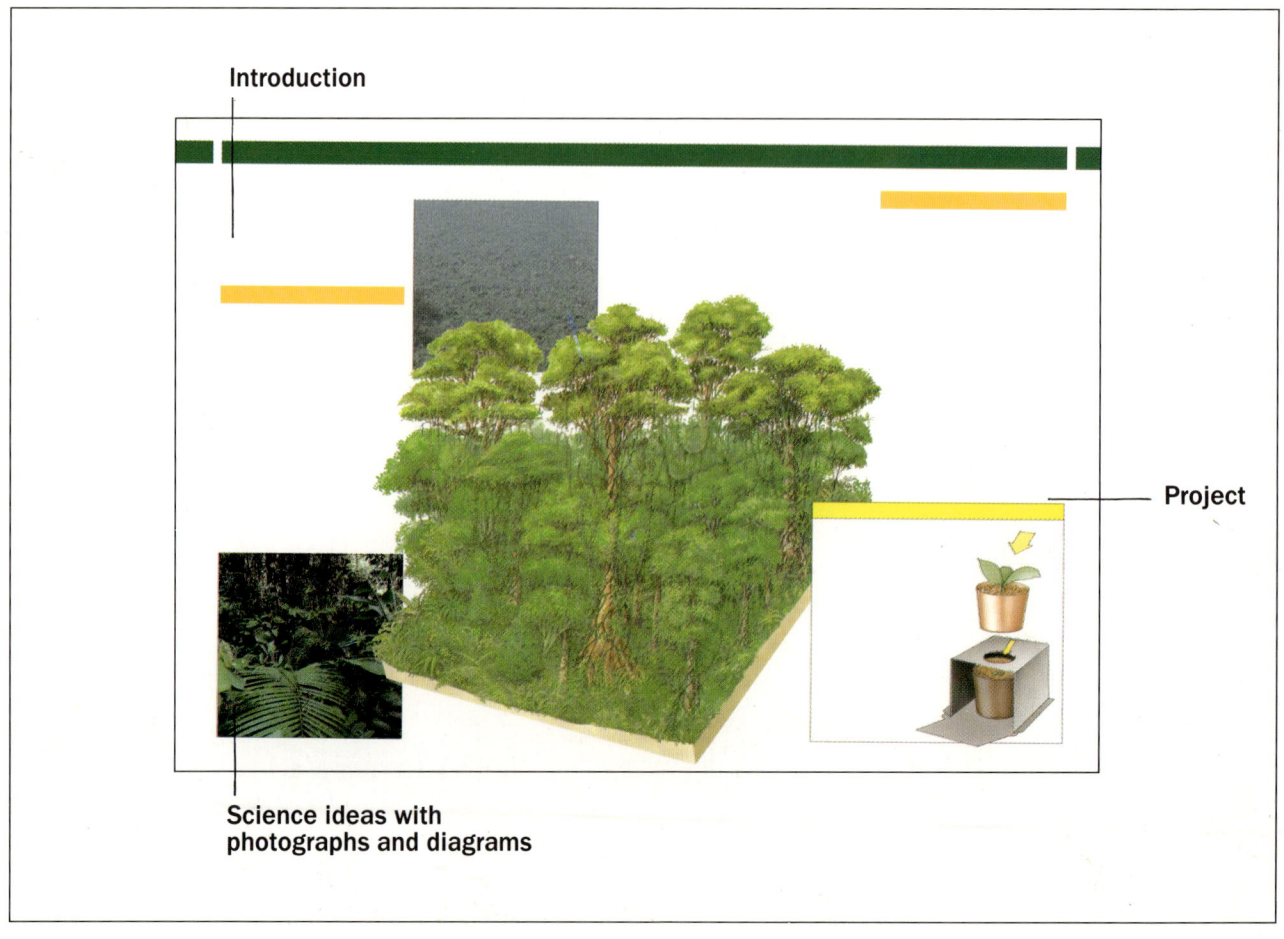

Introduction

Science ideas with photographs and diagrams

Project

INTRODUCTION

5

Forests are an important part of the earth's landscape. They protect the land from erosion, provide homes for millions of plants and animals, and help regulate the oxygen in the atmosphere.

The dominant life-form in forests is trees, and the type of trees that thrive in a particular forest depends on the climate of the area. For example, coniferous trees, which stay green all year around, thrive in the cold climate of the North. Deciduous trees that shed their leaves in winter cover large areas of the temperate forests. Tropical rain forests have a year-round growing season, and contain an amazing diversity of plant and animal life.

Forests compete with human beings for use of the land. Ten thousand years ago, the world's forests were twice as large as they are today. But logging and agricultural activity have reduced much of the forests over the years, and many people are worried about the rate at which forests are being cut down.

Beautiful ancient oak trees in a forest in Dartmoor, England

FOREST ZONES

Today, about one-quarter of the earth's land surface is covered by dense, green forest. Stretching over millions of square miles, trees are the most dominant life-form. The world's forests can be divided into three main types – the tropical rain forests, the temperate forests, and the boreal forests.

WHERE THEY GROW

Forests develop over many years, and different types of plants and animals live in the forest until it reaches its climax stage. For example, after a few years, pines may start to grow in a grassy meadow. These are finally replaced with wholly deciduous trees, the final stage in the forest succession.

The type of trees that grow in a particular area depends on the temperature and rainfall. In common with other plants, trees need average temperatures of at least 59°F (15°C) during the summer growing season. Trees also require substantial amounts of rainfall.

Polar regions are too cold for forests to grow.

Deserts are too dry to support dense forests.

BOREAL FORESTS

Boreal forests are the coldest and driest in the world, and are found in areas that have very cold winters and a short growing season. The word boreal means northern, and the forests stretch across Asia, North America, and Europe. High mountain slopes on these continents are also covered with boreal forests.

To the south, boreal forests merge into temperate forests and grasslands. Despite abundant snowfall, boreal forests suffer from drought. Conifers and a few hardy trees such as birch and some willows, are the only trees that can withstand these cold, dry conditions.

◁ Coniferous trees such as spruce and pine retain their needles (leaves) throughout the year. Instead of bearing fruits, the trees bear cones.

TEMPERATE FORESTS

Temperate forests grow in areas with moderate climates. Most temperate regions have warm summers and cool winters, and trees vary from broadleaf, deciduous trees to evergreen trees.

In the Northern Hemisphere, temperate deciduous forests extend across eastern North America, western Europe, and eastern Asia. Temperate evergreen forests are found in western North America.

In the Southern Hemisphere, evergreen forests occur in southeastern Australia and New Zealand. These forests lie on steep slopes that have very high rainfall.

▽ Tropical trees tend to have very tall trunks with the foliage (leaves) right at the top of the tree. The trees normally support lianas (creepers).

△ Temperate, deciduous trees have a full, bushy shape. In autumn, the trees shed their leaves before they can be damaged by frost.

TROPICAL FORESTS

Tropical forests include both dryland forests, in much of east Africa, and rain forests. Tropical rain forests are found in South and Central America, west and central Africa, and all over Southeast Asia and tropical Australasia.

The tropical rain forest zone is characterized by an abundance of rainfall and very high temperatures. These conditions have made them the most spectacular forests on earth today. Tropical rain forests consist mainly of broad-leaved trees that do not shed their leaves annually, but have a year-round growing season.

8 NATURAL COVER

Forests form the natural vegetation for a large part of the earth's land surface. Ten thousand years ago, forests were more extensive than today. For example, western Europe was covered by an unbroken chain of trees. Since then, increasing human activity has drastically reduced the size of the earth's forests.

PEOPLE VERSUS TREES

Farmers have a difficult relationship with trees. While other people may think that trees are a valuable natural resource, farmers sometimes feel land that the trees occupy could be used to plant crops and raise cows and sheep.

Trees are incredibly useful. Wood is a superb construction material. The same tree that provides strong lumber for building bridges also provides delicate struts for tables and chairs. Humans also discovered other uses for wood, most importantly that of making paper. The newspapers, magazines, notebooks, and envelopes we use today are all made from trees.

Trees are also a very important source of fuel. Billions of people all over the world still rely on firewood for everyday cooking and heating.

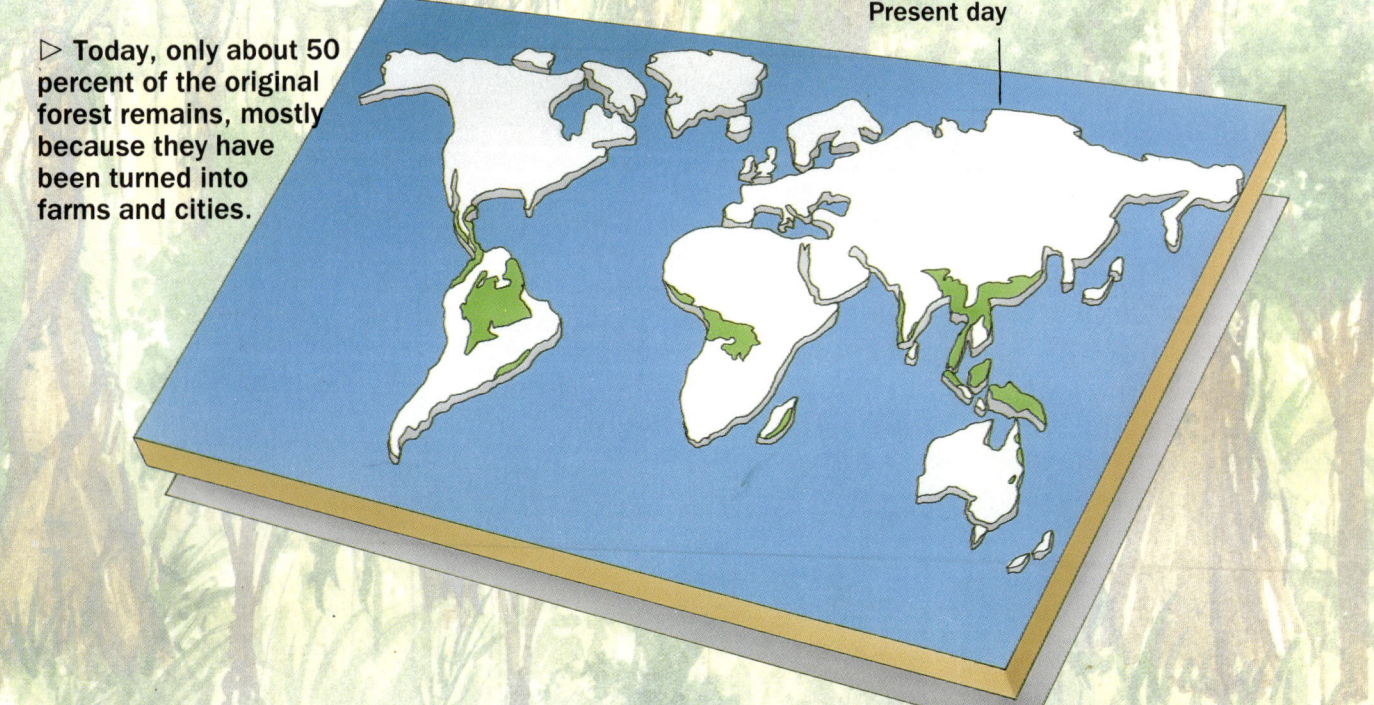

△ The natural forest zones — temperate, boreal, and tropical — covered a large part of the Earth's land surface some 10,000 years ago.

▷ Today, only about 50 percent of the original forest remains, mostly because they have been turned into farms and cities.

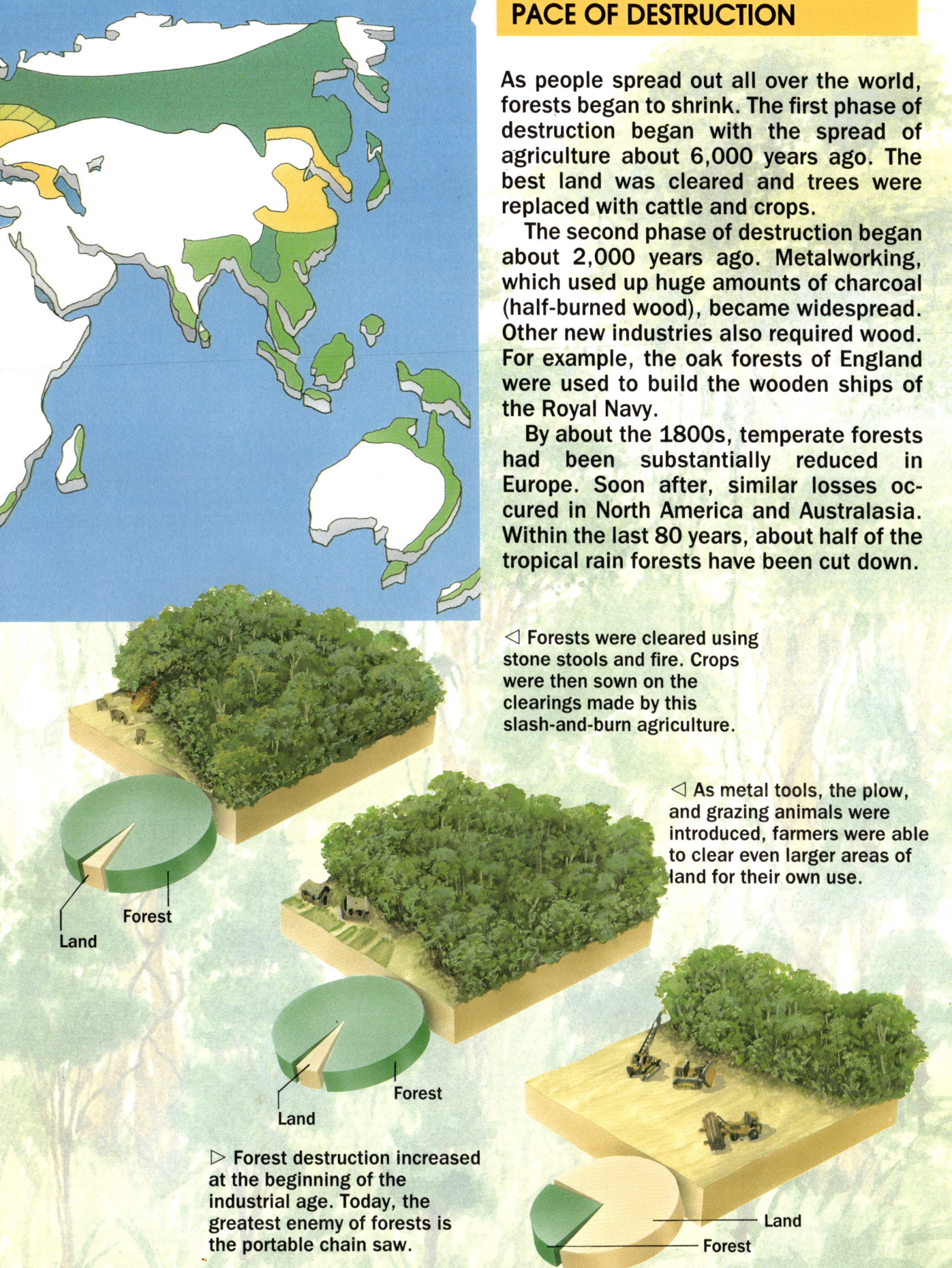

PACE OF DESTRUCTION

As people spread out all over the world, forests began to shrink. The first phase of destruction began with the spread of agriculture about 6,000 years ago. The best land was cleared and trees were replaced with cattle and crops.

The second phase of destruction began about 2,000 years ago. Metalworking, which used up huge amounts of charcoal (half-burned wood), became widespread. Other new industries also required wood. For example, the oak forests of England were used to build the wooden ships of the Royal Navy.

By about the 1800s, temperate forests had been substantially reduced in Europe. Soon after, similar losses occured in North America and Australasia. Within the last 80 years, about half of the tropical rain forests have been cut down.

◁ Forests were cleared using stone stools and fire. Crops were then sown on the clearings made by this slash-and-burn agriculture.

◁ As metal tools, the plow, and grazing animals were introduced, farmers were able to clear even larger areas of land for their own use.

▷ Forest destruction increased at the beginning of the industrial age. Today, the greatest enemy of forests is the portable chain saw.

TROPICAL RAIN FORESTS

Because of the favorable climate around the equator, tropical rain forests have become the most complex and diverse places on Earth. These forests contain the greatest variety of plant and animal species. But due to unrestricted logging, these forests have also become the most threatened places on Earth.

EXTENT

At the beginning of the 20th century, there were about 6 million square miles of tropical forest. Today, less than 3 million square miles remain.

The destruction of tropical forests began with large-scale logging operations. Many tropical species, such as teak and mahogany, were felled because they provided valuable lumber for humans. However, such logging is very inefficient. Because the forests are so dense and diverse, about 50 other trees are cut down for every lumber tree.

When forests are cleared for agricultural purposes, the trees are often burned down. Huge areas have been cleared in this way for agriculture and cattle ranching, and during the 1980s, tropical forests were being cut down at a rate of 2 percent a year.

Petrified logs found in Arizona

Palm trees in Brazilian rain forest

HISTORY

Tropical forests are a product of their climate, and this has not always been as it is today. Between 400,000 and 14,000 years ago, our planet experienced a series of Ice Ages, when the climate was much colder and drier. As the climate got colder, the tropical forests shrank.

At the height of the Ice Ages, there were probably no tropical forests in South America or Asia. However, in Africa, conditions were better. Scientists have identified areas of forest in west Africa which they believe existed through the Ice Ages. This survival means that they are the oldest on Earth.

When the climate became warmer the tropical forests began to grow again. They probably reached their maximum extent just before humans commenced their large-scale destruction.

LOCATION

Tropical forests grow well around the equator where temperatures rarely rise above 95°F or drop below 68°F. The forests also receive about 80 inches of rain a year. The air beneath the trees is always humid, and in areas where sunlight reaches the ground, dense growth called jungles occur. Most of these are found near rivers.

The largest area of tropical forest in the world occurs in the vast Amazon River Basin of South America. Smaller stretches extend north to the West Indies. In Africa, the forest follows a similar pattern, and is mainly confined to the basin of the Zaire River.

In Asia, the tropical forest zone is more widespread. In eastern Asia, the natural forest extends from southern China to northern Australia, and covers most of the islands in between. In western Asia, there is a narrow rain forest zone down the west coast of India.

Part of the Zaire rain forest in Africa

CREATE TROPICAL CONDITIONS

Plant two seedlings in pots. Place one in a cool place. Cover the other with a transparent plastic bag, and keep it warm. Give both seedlings equal light and water. After a week, see which one has grown the most.

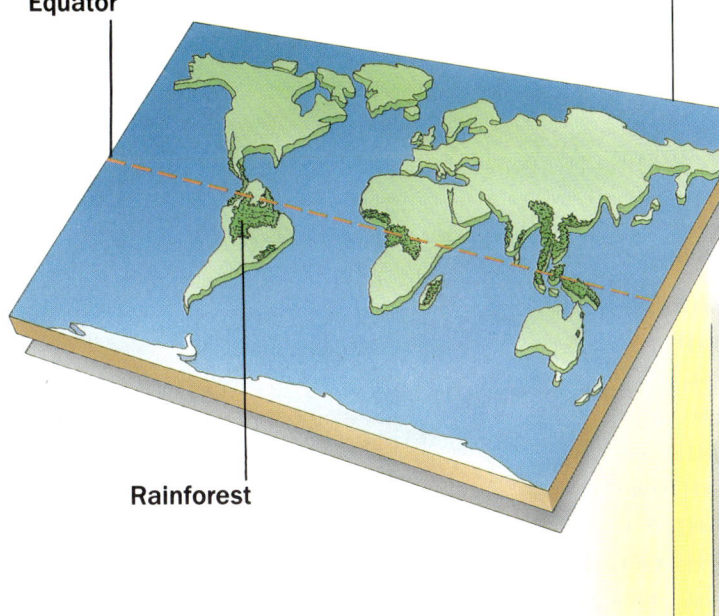

△ The tropical rain forest zone occurs around the equator in South America, Africa, and Southeast Asia. Rain forests also cover parts of Australia.

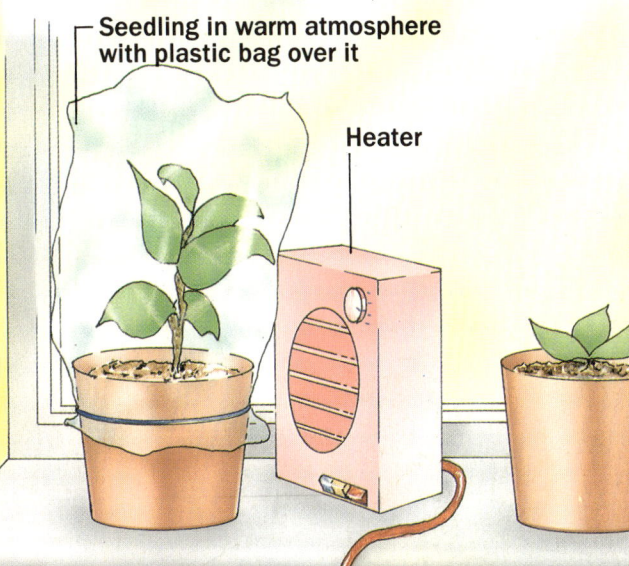

12 LAYERED FORESTS

Tropical rain forests form a multilayered environment. Contrary to popular opinion, not all rain forests contain the dense, tangled undergrowth that is known as jungle, although most of them have a dense growth of trees. The most crowded part of a tropical rain forest is high above the ground, in the treetops.

LIFE AT THE TOP

A tropical rain forest can be divided into three distinct layers, one above the other. About 130 feet above the ground, the leaves and branches of the trees form the first forest canopy. The trees grow so close together that their branches often intertwine. Occasionally a taller tree, known as an emergent, will tower an additional 33 to 65 feet above the rest of the canopy.

A second canopy normally reaches a height of 65 feet, and the third reaches 33 feet. The shrub and herb layers of tropical rain forests are quite thin. This is because very little sunlight penetrates to the forest floor. The canopies also have many climbing plants and epiphytes growing in the branches.

Aerial view of rain forest in Venezuela

Rain forest vegetation in Surinam

Undergrowth

▽ Tropical rain forests have three distinct canopies. Tall trees, called emergents, sometimes grow higher than the first canopy. The undergrowth is normally thin.

Emergents

Gaps

Canopy

GROWTH FACTOR

Tropical rain forests need plenty of sunlight for them to survive. The tall trees in the canopies use their green leaves and the sunlight to make food. This process is known as photosynthesis. There is a lot of competition for sunlight, and this makes the canopy quite crowded. Beneath the canopy, trees cast a very dense shade. Less than 2 percent of the sunlight reaching the canopy filters down to the forest floor. Few plants can grow in such darkness, and the warm, moist conditions mean that the rate of decomposition is extremely rapid.

A natural forest canopy has some spaces within it. When a large tree dies it collapses, often bringing down other trees and producing a small clearing. At any given time, about ten percent of the forest consists of such clearings. Here, sunlight penetrates to the ground, and the undergrowth becomes dense. Eventually, a few trees grow to reach the canopy, and the clearing disappears.

RESTRICTED LIGHT

You can demonstrate the effects of the canopy with two seedlings. Plant one seedling high in its pot, and place it on a windowsill. Plant the other seedling low in its pot, and construct a cardboard shield as shown in the picture. Place the shielded pot alongside the other one, and give both the same amount of water. Measure the seedlings every few days, and keep a record of how much each one has grown.

14 DIVERSITY

Tropical rain forests contain plant and animal life at its most abundant. We will never know how many species the forests originally contained, because species are becoming extinct faster than we can count them. A safe estimate is that they now contain at least one million different species, most of which are insects.

VARIETY OF LIFE

In terms of plant species, the richest forests are those of Southeast Asia. The Malay Peninsula contains more than 2,500 tree species, including teak and palm. A single square mile of tropical rain forest may contain more than 100 different types of trees and hundreds more smaller plants. By comparison, other forests have far fewer species.

This great diversity of plant life and the layered structure of the forests have given rise to an even greater diversity of animal life. A single tree might easily support hundreds of different insect species and a variety of animals. Most animals spend their lives in the trees, feeding on the nuts and fruits in the upper two canopies.

Rain forest epiphyte

Strangler figs

Orange orchid

Fungi

Bromeliads growing on fallen tree trunk

▽ Tropical rain forests are the most diverse forests on Earth. In South America, the forests contain more than 70 different species of tree in just one acre.

- Tapang
- Arial pitcher
- Fern

WAYS OF LIFE

Tall trees dominate the forests, taking up most of the available space, light, and nutrients. Competition between smaller plants is intense, and forest trees support numerous epiphytes and parasites.

Epiphytes are plants that grow on tree branches without harming them. As leaf litter decays to form a thin soil, many plants take root there. Typical epiphytes are bromeliads, which have large, cup-shaped flowers that are able to collect rainwater.

Parasitic plants send out roots to absorb water and nutrients from the tissues of a host plant. Some plants, such as the strangler fig, start out harmless, but later turn parasitic, enveloping the host completely and killing it.

Many saprophytes live on the forest floor. Saprophytes are plants that absorb nutrients from decaying organic matter. Some tropical orchids and many types of fungi are saprophytes. Bacteria and other decomposers on the ground speed up the process of decay.

Orchid in Peru's tropical rain forest

HOW IT WORKS

As their name suggests, tropical rain forests depend upon a high amount of rainfall. This abundant water is the key to the lushness of the forests. At the same time, the forests also protect the earth from the damaging effects of the rain and wind. Without the forests, the land would rapidly turn into a wasteland.

RAINFALL

Rain forests are generally found where the rainfall is higher than 60 inches a year and the dry season is no longer than two months. An exception to this rule is found in parts of Southeast Asia, where most of the rainfall is concentrated in the annual monsoon.

Heavy rainfall is a result of global climate patterns. However, rain forests also contribute to the amount of moisture in the atmosphere. Some of the rain that falls evaporates from the tree canopy. The rest is absorbed by the roots and transported to the leaves. The leaves give off water through their pores, and this process is known as transpiration.

Above the forest, the evaporated water condenses to form thick clouds that produce brief thunderstorms on an almost daily basis.

Monsoon rain falling on the Ujung river in Indonesia

▽ Although rain forests recycle much of the rainfall back into the atmosphere, a considerable amount of water seeps away into the soil.

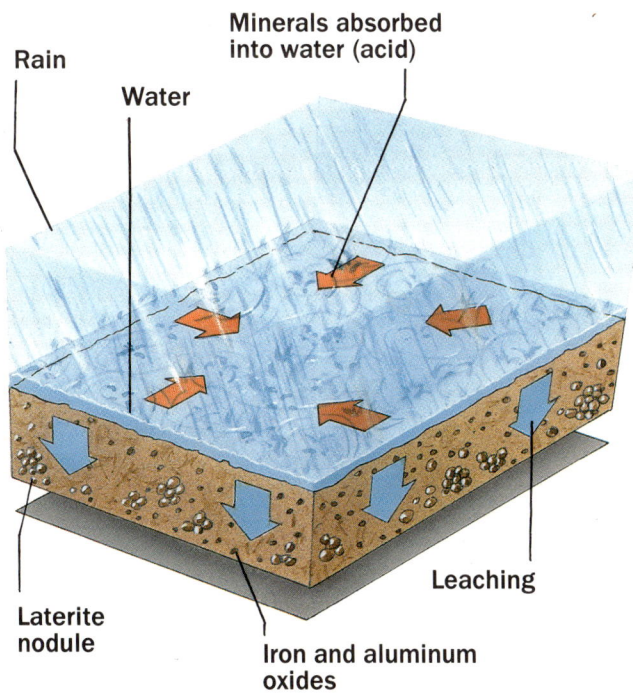

▽ Without the protection of trees, tropical soils deteriorate very rapidly. Within a few years the soil becomes filled with mineral nodules that turn into laterite.

LEACHING PROJECT

Half fill a flowerpot with gravel and sprinkle on some sand. Now add a layer of salt crystals, and cover with 1 inch of sand. Water the pot every day. See how much salt is left after one day; one week.

SOILS

Surprisingly, tropical forest soils are relatively poor and infertile. About 95 percent of the available nutrients are locked into the tissues of the giant trees.

Because of the rapid rate of decomposition, there is only a very thin layer of decomposing organic matter on the forest floor. Any nutrients released are either absorbed by the trees or washed down through the soil by rainwater. This process is known as leaching.

In tropical regions, leaching also makes certain minerals (notably iron and aluminum) rise to the surface. These minerals turn the soil acidic, inhibiting plant growth.

When trees are cut down, the process of leaching speeds up because there is less evaporation. The action of sunlight on the exposed soil often turns it into a useless red dust called laterite. Laterite is easily blown away by the wind.

TEMPERATE FORESTS

Most temperate forests lie between hot, wet tropical forests and cold, dry boreal forests. Their most important feature is their ability to adapt to seasonal changes in climate. During autumn, they lose their green color when the trees shed their leaves, and by winter they seem bare and lifeless.

LOCATION

Temperate forests are mainly confined to the Northern Hemisphere and occur in Europe, northern Asia, and North America. Trees need at least 28 inches of rainfall per year, so areas with less rainfall are unable to support this type of forest. Temperate forests tend to occur within about 600 miles of the coast. Further inland, the weather is generally too dry to support them.

Temperate trees grow much farther apart than those in tropical forests. The leaves and branches of adjacent trees are not as tightly packed, and a reasonable amount of sunlight is able to reach the forest floor.

The sunlight that reaches the ground encourages growth of a large number of other forest plants, mainly herbs and small shrubs. The undergrowth in some places can become quite thick and tangled. Near the forest floor conditions are often humid, and this encourages ferns and fungi.

Woodland scene

▽ An example of the type of plant life that grows on a temperate forest floor: (1) wild mushrooms, (2) ferns, (3) shrubs, and (4) undergrowth.

DECIDUOUS HABIT

In temperate regions, the winters are short but often severe. Short days bring a sharp reduction in the amount of heat and light. When temperatures fall below freezing, very little rain falls on the trees and delicate plant tissues may be fatally damaged by frost.

Many small forest plants die away completely at the onset of winter and grow again from seed the following year. Trees, however, are much longer-lived and have to survive the harsh conditions. Most temperate trees have adapted themselves by becoming deciduous — shedding their leaves in autumn and virtually shutting down for the winter. Without leaves, there is no photosynthesis and the tree becomes dormant. In spring, leaf buds appear on the branches and the forest grows again.

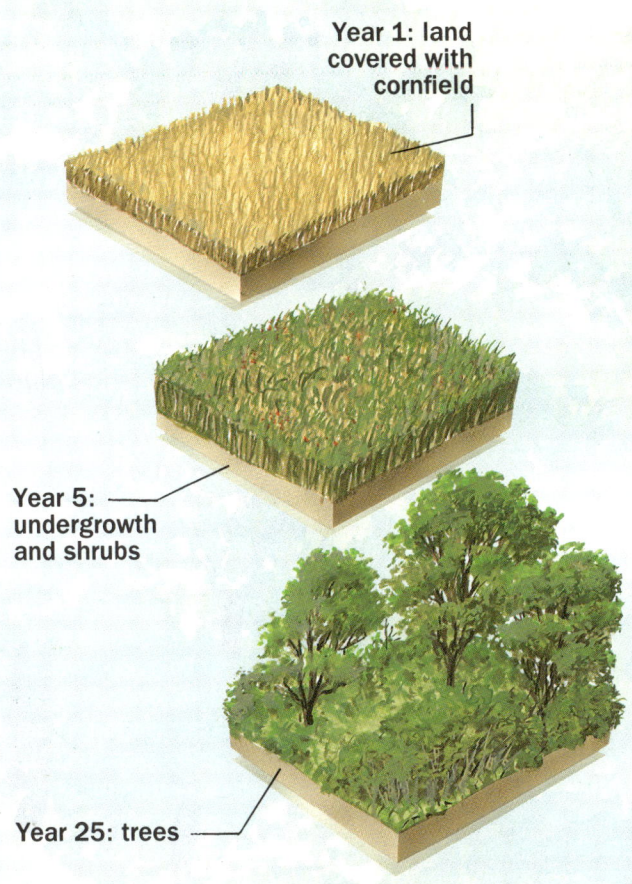

Year 1: land covered with cornfield
Year 5: undergrowth and shrubs
Year 25: trees

FOREST REGROWTH

Throughout history, huge areas of temperate forest have been cleared for agricultural purposes. Occasionally, fields have been abandoned and the natural vegetation allowed to regrow. Scientists have been able to observe the stages of succession — the various types of vegetation that grow before the climax community is established.

The first plants to invade the abandoned fields were wild grasses and herbs, including most common weeds. These provided sufficient shelter for woody plants and shrubs to take hold. Trailing plants such as wild roses, blackberries, and ivy covered the ground. After about 25 years, young broad-leaved trees became well established.

◁ The annual deciduous cycle.
1) Spring: leaves grow. 2) Summer: tree is in full leaf. 3) Autumn: leaves dry out and fall. 4) Winter: most trees are bare.

20 SPECIES AND SOIL

Temperate forests are much less diverse than tropical forests, and the number of species per square mile is measured in tens, rather than hundreds. Temperate forests usually contain a mixture of tree species, but local weather conditions can lead to one particular species becoming dominant.

SPECIES

Typical trees of the European temperate forests include oak, ash, elm, beech, horse chestnut, lime, willow, and hazel. North American and north Asian forests contain a wider range of species, including magnolias, maples, and hickories.

In cooler, drier areas, temperate forests often have abundant numbers of silver birch trees. In places with poor climate and soil, coniferous trees which are able to resist the harsh climate thrive.

The temperate forest zone once covered almost all of the Northern Hemisphere. Over the centuries, most of the trees were cut down to provide lumber and fuel, and make room for farms and cities. Humans also have some part to play in the present species distribution.

For example, the sycamore is not native to the British Isles, but was probably introduced by the Romans. In areas with mild winters, evergreen species are able to develop. Their leaves have a waxy coating that helps them withstand winter drought. The evergreen rhododendron, native to the Himalayas but now widespread, can survive winter beneath an insulating blanket of snow.

Bamboo forest in China

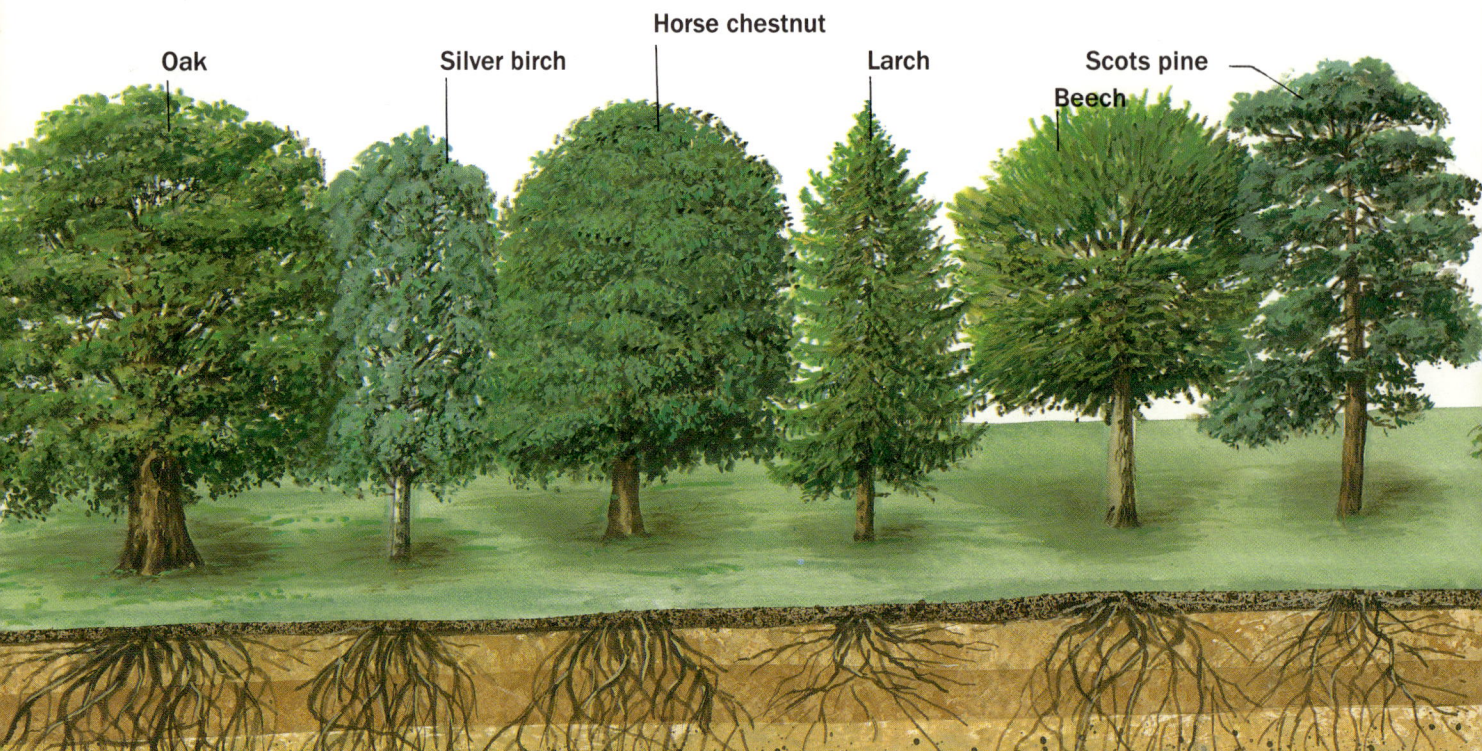

SHADE AND SOIL

Trees vary in the amount of light they require, and in the amount of shade they cast. Beech trees absorb the most amount of sunlight. For every square yard of land beneath a beech tree, there may be almost eight square yards of beech leaves above. As a result, beech trees cast the densest shade, and few other species can survive under them. Beech forests are usually clear of undergrowth.

Shade from forest trees is also responsible for the timing of the first flowers of spring. Plants such as the bluebell have to complete their annual cycle of flowering and fruiting before the shade becomes too dense.

In autumn, the forest floor becomes covered with a carpet of fallen leaves. These leaves build up to form a thick layer of decomposing leaf litter, which eventually rots to form humus and soil.

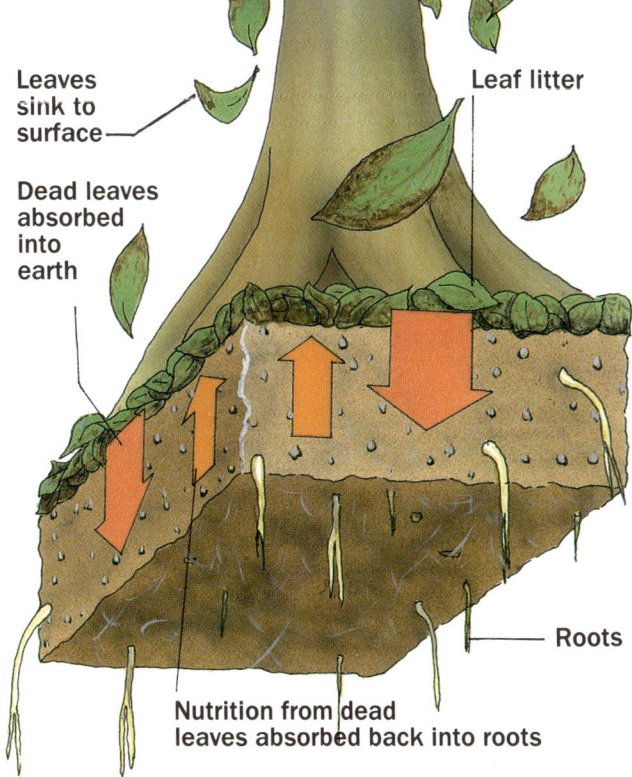

△ Dead leaves that fall to the forest floor add nutrients to the soil. These are absorbed by the trees.

Shedding leaves is also a way of storing nutrients until the following year. During winter, the low temperatures slow down decomposition and nutrients are preserved until the next growing season.

Small animals such as earthworms and insects mix the decomposed leaves with the soil. Nutrients released into the soil are absorbed by the tree roots, and over a long period of time, temperate forests develop very fertile soils.

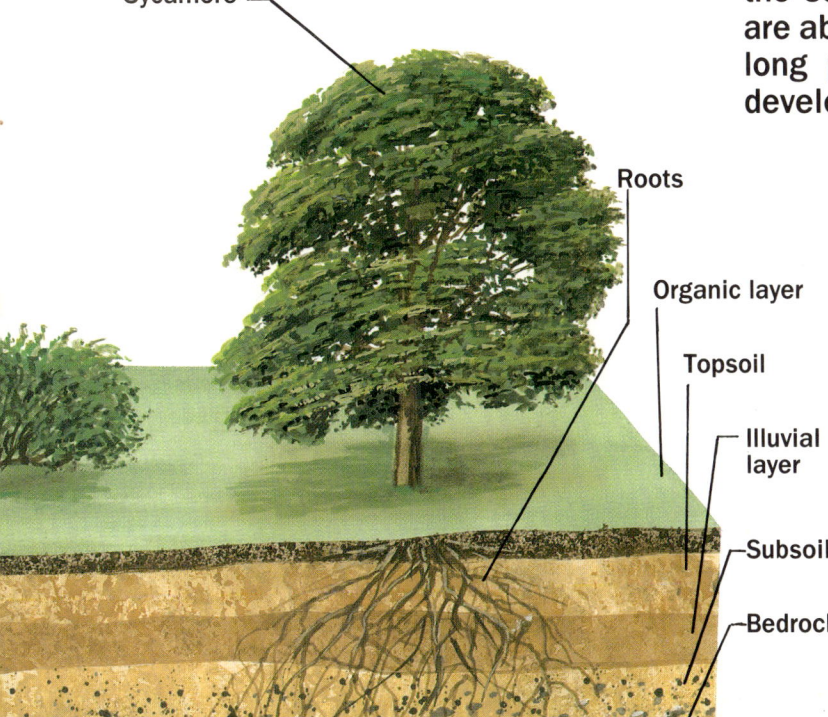

▽ Temperate forests create fertile, well-mixed soils as tree roots gradually break up the bedrock. Many types of trees, from old oaks to horse chestnuts and sycamores, grow in the forests.

Bluebells growing on the clear floor of a beechwood forest

BOREAL FORESTS

Boreal forests are the cold, coniferous evergreen forests of the Northern Hemisphere. Most of them contain conifers, an ancient group of trees that evolved long before broad-leaved trees and other flowering plants. However, the present-day boreal forests are the youngest forests on Earth.

Coniferous trees are extremely hardy and can tolerate the cold northern winters. They are also well-suited to the poor clay and sandy soils left behind by glaciers. Once established, conifers tend to dominate the landscape because they withstand conditions that most other plants cannot tolerate.

EXTENT

Boreal forests form a wide belt around the top of the globe. In Eurasia, the forest is often known by its Russian name, taiga. To the north, boreal forests are limited by the tree line. Beyond this, the polar regions are too cold for any trees except dwarfs a few inches high. To the south, boreal forests merge into temperate forests and grasslands.

Most boreal forests grow on areas of land that were once covered by ice sheets. As the glaciers retreated, coniferous trees occupied the newly exposed landscape. In some places, such as Alaska, boreal forests cover areas that were glaciated only 200 years ago.

Boreal forest is the most northerly of the forest zones.

▽ The boreal forest zone covers much of the northern Arctic regions. Most of the trees in the forest stay green throughout the year.

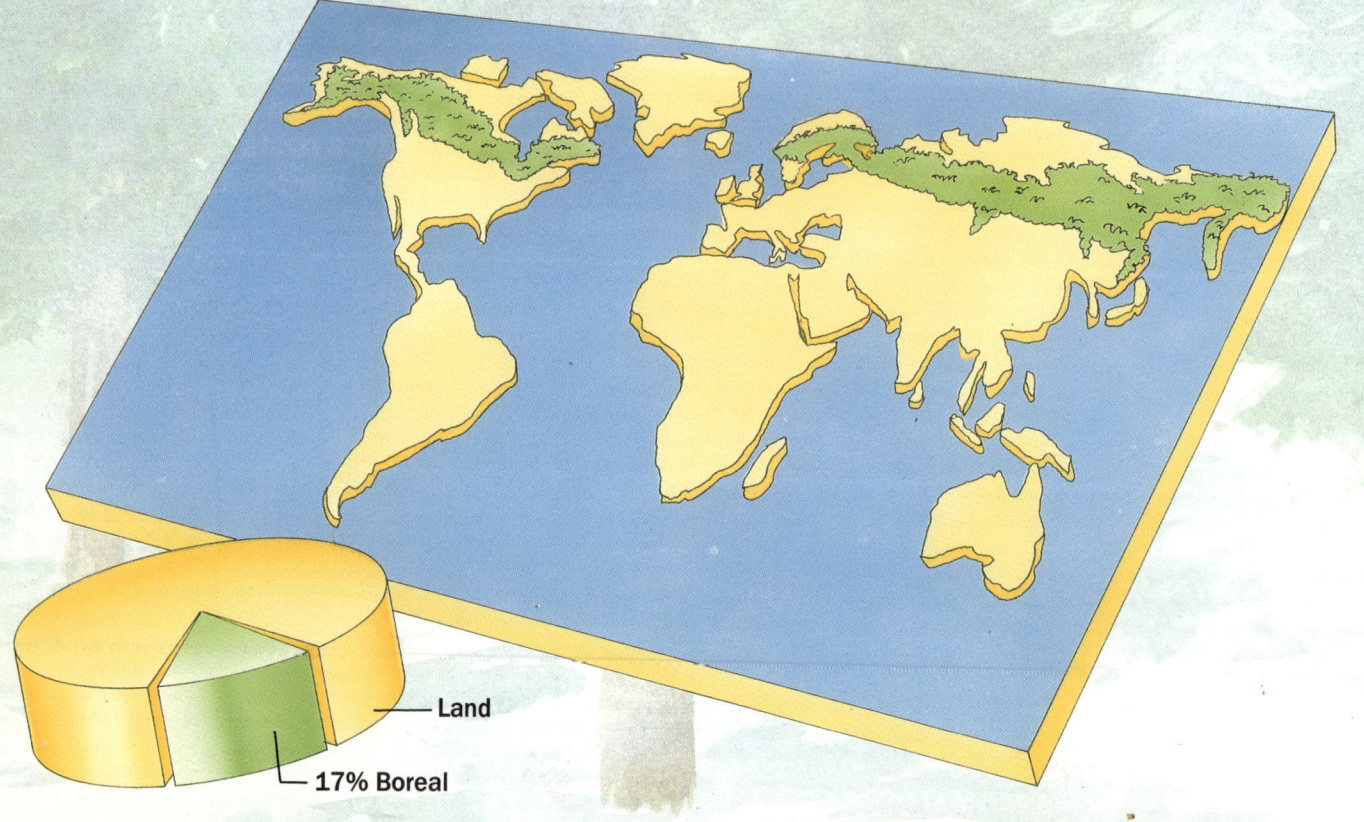

Land

17% Boreal

SPECIES

Boreal forests consist mainly of the various species of fir, spruce, and pine. In drier areas, and especially in Siberia, larches are the dominant species. Boreal forests are less diverse than other forests, and although areas of mixed species do occur, it is more usual to find the same tree species covering hundreds of square miles. This uniformity is characteristic of boreal forests.

Coniferous trees grow much closer together than broad-leaved trees. Spruce, for example, often grows as densely as 12 trees per 100 square yards. This close-packing creates extremely thick shade, with no light for other plants. Spruce forests are usually entirely clear of undergrowth. Pine forests, however, are much less shady and many grasses grow on the forest floor.

Conifers are not the only trees in boreal forests. Around lakes and clearings, hardy trees can be found. But they are soon smothered by the conifers.

A boreal forest in Norway

▽ Trees in a boreal forest belong to several coniferous families. Conifers make up 15% of the world's tree species.

— Scots pine
— Balsam fir
— Larch
— White spruce
— Other species
— Conifers

DID YOU KNOW?

Coniferous trees are the tallest and largest living organisms. One redwood tree in northern California, measures over 367 feet high. A specimen of the giant sequoia species is shorter but thicker, and is the heaviest living thing, weighing more than 2,100 tons.

24 WINTER SURVIVAL

The trees of boreal forests have to endure very harsh winters. Most of the time the temperatures remain below freezing point, and may fall as low as −40°F. Winter also brings severe drought to the area, and groundwater becomes totally frozen. The land ends up being covered in snow, with water in very short supply.

SURVIVAL STRATEGIES

Nearly all conifers retain their needles (leaves) throughout the year, so that they can make their own food. Individual needles have a small surface area and are coated with wax. These adaptations help prevent water loss through evaporation. Natural antifreeze produced in the needles prevents them from being killed by frost.

In the coldest, driest regions, larches lose their needles in winter to keep water loss to a minimum.

Coniferous trees have shallow, spreading root systems that collect water from a wide area. These roots enable conifers to take advantage of the available nutrients during early spring, when only the top few inches of soil thaw out.

The close packing that creates such dense shade also helps keep the trees warm. Air trapped between needles and branches forms a layer of insulation around each tree. The trees' conical shape allows snow to slide down before its weight breaks the branches.

▽ Conifers make maximum use of winter sunlight. Their shape allows some sunlight to reach each tree. Heat reflected up from fallen snow is trapped beneath the trees.

Conifers covered in winter snow

NEEDLES AND SOIL

Dead needles fall to the ground at a steady rate. Over many years, these needles build up to form a thick carpet. Because of the cold climate, needles decompose very slowly. Acidic chemicals in the needles also deter decomposing organisms in the soil.

Dead needles contain few nutrients. Their acidic chemicals discourage earthworms and other burrowing animals from mixing them into the soil. The result is an acidic, infertile soil, unsuitable for plants other than conifers. By affecting the soil in this way, conifers increase their own domination of the landscape.

Not even conifers can flourish in these conditions without assistance. Most owe their success to mycorrhizalic fungi that live in the soil. These fungi supply the tree roots with vital minerals and receive food in exchange. Mycorrhizal fungi are found associated with many trees in all types of climates.

A pinecone from a conifer

▽ Coniferous trees produce infertile soils. The shallow root systems of the trees are able to collect water during the spring thaw.

- Permafrost
- Mineral rich layer
- Pine needles

26 OTHER FORESTS

There are several other forest regions that do not fit the main forest zones. These forests owe their existence to a variety of local climatic conditions. There are warm, dry forests; cool, damp forests; and even saltwater forests. Together they demonstrate how adaptable and widespread trees have become.

ALTITUDE

The temperature of the atmosphere drops by about 1°F for every 1,000 feet above sea level. In the tropics, mountains provide areas of cool weather in an otherwise hot climate. The sides of a mountain can be graded into a series of vegetation zones. At the top of the mountain, it is too cold and icy for plant life. Further down, as conditions get warmer and wetter, herbaceous plants and trees start to appear. Conifers are often found on the higher slopes, with temperate species growing farther down. Cloud forests occur where bands of permanent low cloud create cold, damp conditions.

Mountain forests are not confined to the tropics. In North America, the Rocky Mountains enable boreal species to grow much farther south than they could on low-lying land.

Alpine meadow in France during summer

▽ Bands of vegetation on a mountain. (1) Snow and ice, (2) Alpine meadow, (3) Coniferous forest, (4) Deciduous forest, (5) Tropical forest, (6) Deciduous forest, (7) Coniferous forest, and (8) Tundra meadows.

Mountain forest in Nepal

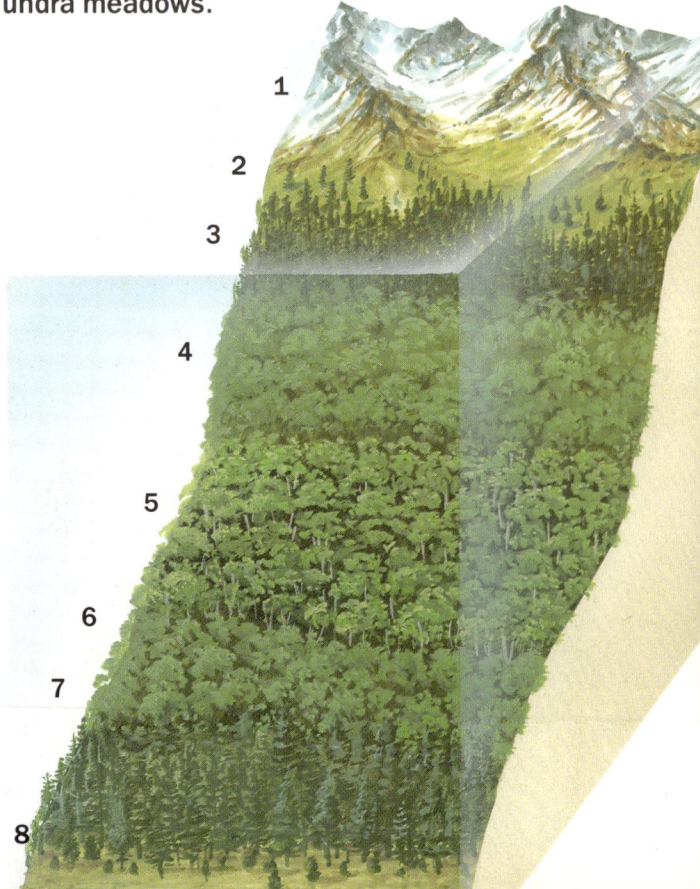

ARID FORESTS

Around the shores of the Mediterranean Sea, there once was a belt of widely-spaced trees that could withstand the hot summers. Much of this forest was cut down to provide farmland for vineyards and olive groves. In areas that have been abandoned, the landscape is now covered with a dense layer of low evergreen shrubs. Some extensive forest areas remain. One of the most unusual trees of the Mediterranean forest is the evergreen cork oak. It is rarely found growing wild, and is extensively cultivated for its bark which is the world's major source of natural cork.

Around tropical grasslands, similar areas of open woodland can be found. In Africa, the dominant species are the acacias. Well adapted to arid conditions, acacias are often covered with thorns to discourage grazing animals.

A Mediterranean forest in May

Waipoura Kauri forest in New Zealand

COASTAL FORESTS

In New Zealand, Canada, and the northwestern United States, climatic conditions have produced temperate rain forests on the slopes of coastal mountains. Warm, damp air from the sea ensures a high rainfall that supports very lush forests. In New Zealand, the temperate rain forests are notable for their many species of tree ferns. American forests contain more conventional temperate species, and many trees are covered with a thick blanket of moss.

Along many tropical and subtropical coastlines are narrow belts of mangrove swamp. Mangrove trees can tolerate salty water, and some species grow out into tidal waters. Arched aerial roots enable the trees to survive the waterlogged conditions. Farther inland, where the coastal swamps are less salty, other water-tolerant species, such as swamp cypress, may be found.

28 FOREST MANAGEMENT

Deforestation, the loss of natural forest, is now the focus of worldwide attention. Tropical rain forests are being cut down at the rate of about one square mile every eight minutes. Many countries have now placed strict controls on the use of their forests, but in many cases the damage has already been done.

LOST TREES AND LOST SOIL

Logging, oil drilling, agriculture, human settlement, gold mining, road construction and many other activities all contribute to the continuing pressure on rain forests. In Brazil and India, for example, the coastal rain forests have almost disappeared completely.

Other forest zones suffer from different pressures. In arid regions, extended drought and the human need for firewood combine to threaten the scant remaining forests. In Southeast Asia, vast areas of mangroves have been cut down to make disposable chopsticks.

The effects of deforestation are worse in hilly, tropical areas. Without trees to anchor the soil, the heavy rain soon washes the soil from the hillsides, leaving steep-sided gullies. The soil is washed into rivers, carried downstream and deposited, choking irrigation channels and smothering fields.

A deforested hillside in Nepal, now used for agriculture

A deforested area in Brazil

△ The rate of deforestation is most severe in the tropics. The rain forests of Brazil are being cut down at an alarming rate, and many environmentalists fear for their survival.

TREE COVER PROJECT

Make two mounds of sand and sprinkle one with lettuce seeds. After the seeds germinate, water both mounds each day. Over two weeks, see which mound holds its shape better. Notice what happens to the other mound.

△ Careful planting projects can generate trees in deforested areas. This picture shows the results of reforestation in Nepal. In the background is a deforested area.

PRESERVATION AND REPLANTING

Several countries have now banned logging, and their remaining areas of rain forest are protected by law. Environmental groups have played an important role in making people aware of the need to preserve the rain forests.

Halting the destruction is much easier, however, than replanting the rain forest. In many areas, soil erosion is already so severe that the land can no longer support tall trees. Even where conditions are favorable, regrowth is a very long process. It can take up to 600 years for a rain forest to reach a climax community.

In temperate regions, large areas of former deciduous forest have been replanted with quick-growing coniferous trees. Although these conifers provide a renewable source of lumber, they are not a good option. Coniferous trees turn the soil acidic and can have a harmful effect on drainage patterns. If the conifers are not native to the region, they do not support much wildlife.

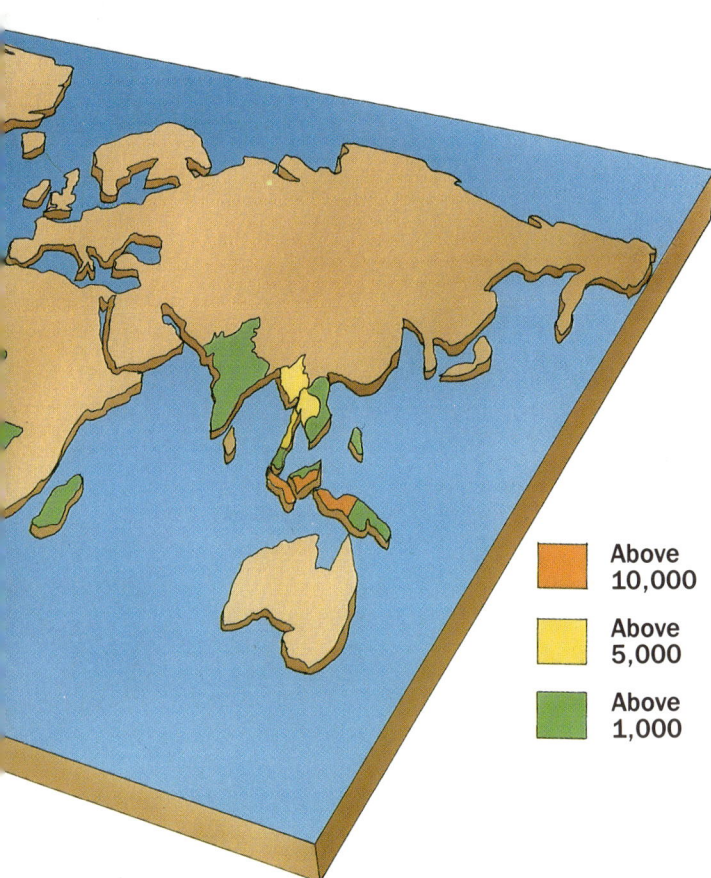

Above 10,000
Above 5,000
Above 1,000

MYTHS OF THE FOREST

Trees are an ancient life-form, and they have existed on our planet much longer than human beings. Individual trees now grow much larger and live for much longer than human beings. The oldest trees are over 2,000 years old. In ancient times, people treated trees with greater respect than they do today.

The Green Man is a pagan spring god associated with the May Day festival.

Throughout history, most people considered forests to be very scary places. The dark, gloomy interior of forests were thought to contain fierce animals, wicked trolls, and evil spirits. In many myths, folktales, and legends, forests feature as strange, uncivilized places where the unexpected can happen. By contrast, forest peoples often regard the forest trees and animals as individuals just like themselves.

A carving from a church in Norway

In ancient India, villagers often selected one particular tree for special attention. As long as the village experienced good fortune, the inhabitants continued to respect the tree. If their fortunes declined, then they would abandon the tree to the wild forest.

Ancient Europe had many tribal religions that paid special attention to trees. Often, trees were planted in sacred rings or groves, which could only be entered by druids and priests. In North America, Ojibway Indians did not like to fell living trees, because they did not want to cause them pain.

In Tolkien's imagination, forest trees became semihuman, with individual personalities.

Forests are a powerful symbol in human art and literature. Few people have not heard a version of "Little Red Riding Hood," who is menaced by a forest wolf. Tolkien's "Lord of the Rings" turns giant trees into creatures called Ents, that can walk around. Much more important are the real forests of the world. They are vital and irreplaceable, providing homes for some of the world's rarest plants and animals.

GLOSSARY

Basin
Large area of land that is drained by a single river system.

Boreal forest
Northern forest zone consisting almost entirely of coniferous trees.

Broad-leaved
Describes trees that have flat, veined leaves.

Canopy
The dense layers of leaves above the floor of a tropical rain forest.

Climax community
The final, relatively stable form of a particular type of vegetation.

Cloud forest
Mountain forest that occurs under a permanent layer of low cloud.

Conifers
Trees with thin needle-shaped leaves that produce seeds in cones.

Deciduous
Describes trees that lose their leaves annually in autumn and regrow them in spring.

Emergent
Tall tropical tree that reaches up to 60 feet above the forest canopy.

Epiphyte
A plant that physically grows on another, but without harming its host.

Evergreen
Trees that retain their leaves throughout the year.

Jungle
Areas of dense, tangled undergrowth rarely found in tropical rain forests.

Laterite
Infertile tropical soil stained red by the presence of iron oxide.

Leaching
The process by which minerals and nutrients are washed out of a soil by rainwater.

Lianas
Trailing tropical creepers.

Mangroves
Trees that grow in coastal subtropical swamps.

Parasite
Plant or animal that grows on, and feeds upon, another organism.

Photosynthesis
The process by which plants use sunlight to convert water and carbon dioxide into food.

Rain forest
Lush forest with a high rainfall. The term is often used to refer to tropical rain forest.

Saprophyte
Plant that does not photosynthesize, but which obtains food from decomposing organic material.

Succession
The process by which natural vegetation goes through a series of stages before it reaches a climax community.

Tundra
Windswept zone with little vegetation found in polar regions.

Understory
The uncrowded middle layer of a tropical rain forest.

INDEX

A
acidic soil 17, 25, 29
agriculture 5, 8, 9, 10, 19, 28
animal life 14, 21
arid forests 27, 28

B
beech forests 21
boreal forests 6, 8, 22-5, 26, 31
broad-leaved trees 7, 19, 31
bromeliads 14, 15

C
canopies 12, 13, 14, 16, 31
climatic conditions 5, 10, 18, 26, 27
climax communities 6, 19, 29, 31
cloud forests 26, 31
coastal forests 27, 28
coniferous trees 5, 6, 20, 22, 23, 24-5, 26, 29, 31

D
deciduous trees 5, 6, 7, 19, 23, 26, 29, 31
decomposition 13, 15, 17, 21, 25
deforestation 8, 9, 10, 20, 28
dryland forests 7

E
emergent trees 12, 13, 31
environmental problems 28-9
epiphytes 12, 14, 15, 31
evergreen trees 7, 20, 22, 31

F
forest floor 12, 13, 15, 17, 18, 21, 23
forest soils 17, 21, 22, 25, 29
forest zones 6-7
fungi 18, 25

J
jungle 12, 31

L
laterite 17, 31
leaching 17, 31
lianas 7, 31
light and shade 13, 21, 23
logging 5, 10, 28, 29
lumber 8, 9, 10, 20, 29

M
mangroves 27, 28, 31
mountains 6, 26, 27
myths and legends 30

N
needles 6, 24, 25

P
parasitic plants 15, 31
photosynthesis 13, 19, 31
pine forests 6, 23
plant life 14, 15, 18, 19, 21, 23
polar regions 6, 22
rainfall 6, 7, 11, 16, 18, 19, 27

R
river basins 11, 31
root systems 24, 25, 27

S
saprophytes 15, 31
soil erosion 29
spruce forests 23

T
taiga 22
temperate forests 6, 7, 8, 9, 18-21
temperate rain forests 27
temperatures 6, 7, 11, 19, 24, 26
transpiration 16
tree species 5, 14, 15, 20, 23
tropical forests 7, 8, 10-17
tropical rain forests 5, 6, 7, 9, 10, 11-17, 28, 29, 31
tundra 31

U
undergrowth 12, 13, 18, 21, 23
understory 31

W
water loss 24

Photographic Credits:
Cover: NHPA; 5, 10 top, 12 bottom, 15, 16-17 & 26 top: Bruce Coleman Ltd; 6 top, 14 top & bottom, 20 & 27 top: Planet Earth Pictures; 6 middle & bottom, 7 top, 18, 21, 22, 23, 24, 25 and 27 bottom: Spectrum Colour Library; 7 bottom, 11, 12 top, 26 bottom, 28 both & 29: Panos Pictures; 10 bottom: Science Photo Library; 30 top left & bottom: Ancient Art and Architecture Collection; 30 top right: With kind permission from George Allen and Unwin (publishers) Limited, of HarperCollins Publishers Limited. Illustration by Alan Lee from THE LORD OF THE RINGS by JR Tolkien.